15 SPACE MISSIONS THAT CHANGED THE WORLD

By Rayan Bale

Your Gateway to the Stars

Copyright © 2024 by Rayan Bale

All rights reserved. No part of this publication may be reproduced, distributed, or transmitted in any form or by any means, including photocopying, recording, or other electronic or mechanical methods, without the prior written permission of the publisher, except in the case of brief quotations embodied in critical reviews and certain other noncommercial uses permitted by copyright law.

TABLE OF CONTENTS

Introduction   3

Mission 1: Sputnik 1:   8
    The first artificial satellite.

Mission 2: Explorer 1:   13
    The first successful U.S. satellite.

Mission 3: Vostok 1:   20
    The first human spaceflight.

Mission 4: Vostok 6:   27
    The first woman in space.

Mission 5: Apollo 8:   33
    The first manned mission to orbit the Moon.

Mission 6: Apollo 11:   40
    The first manned moon landing.

Mission 7: Viking 1:   48
    The first successful mission to land on Mars.

Mission 8: Voyager 1 and 2:   55
    Our first journey to interstellar space.

Mission 9: Hubble Space Telescope: 62
Revolutionizing our understanding of the universe.

Mission 10: Cassini-Huygens: 69
Exploring Saturn and its moons.

Mission 11: International Space Station (ISS): 75
Continuous human presence and research in space

Mission 12: Mars Exploration Rovers: Spirit and Opportunity: 81
Provided evidence of past water on Mars.

Mission 13: Kepler Space Telescope: 86
Discovered thousands of exoplanets

Mission 14: Perseverance Rover: 92
Landed on Mars to find past life and collect samples

Mission 15: James Webb Space Telescope: 99
Studies the early universe and exoplanets

Conclusion 105

References 109

INTRODUCTION

Space exploration is one of the most awe-inspiring and ambitious endeavors humanity has ever undertaken. This book, "15 Space Missions That Changed the World," not only showcases the most significant missions in the history of space exploration but also reveals the remarkable evolution of our journey into the cosmos. Each mission represents a leap forward in our understanding and capability, marking milestones in human ingenuity and the quest for knowledge.

We begin with the launch of the first artificial body in space, marking the start of the space age and igniting a fervent race to explore beyond our planet. This historic achievement set the stage for all future space endeavors, demonstrating that humanity could reach beyond the confines of Earth. From this point, space missions quickly progressed to include manned flights, breaking new ground with each successful mission. The bravery of astronauts who ventured into the unknown opened the door to an era where humans could live and work in space, even if only for short periods at first.

The early years of space exploration were marked by incredible achievements, including sending the first humans into orbit and landing on the moon. These milestones were not just technical triumphs but also profound moments of human achievement that captured the imagination of people around the world. The sight of astronauts walking on the moon for the first time was a testament to human ingenuity and the spirit of exploration that drives us to push the boundaries of what is possible.

As technology advanced, so did our ambitions. Robotic explorers were sent to distant planets, transforming our understanding of the solar system. These missions provided detailed data and images that were previously unimaginable, revealing the complexities and wonders of other worlds. The exploration of Mars, for instance, has been particularly groundbreaking, with rovers uncovering evidence of water and the potential for life beyond Earth.

In addition to these robotic explorers, the development and deployment of space probes to the outer planets have vastly expanded our knowledge of the more distant realms of our solar system. These probes have flown by and studied the giant planets and their moons, unveiling astonishing details about their atmospheres, surfaces, and potential for harboring life. Missions to Jupiter, Saturn, and beyond have revealed the dynamic and diverse nature of these celestial bodies, capturing imaginations and driving further scientific inquiries.

The continuous human presence in space, exemplified by the International Space Station, represents another significant leap. This orbiting laboratory has become a hub for scientific research and international cooperation, providing invaluable insights into the effects of long-term space travel on the human body and advancing our preparation for future deep-space missions.

Our quest to understand the universe also led to the development of powerful space telescopes that have peered deeper into the cosmos than ever before. These observatories have captured stunning images of distant galaxies, nebulae, and stars, enhancing our understanding of the universe's origins and structure. The discoveries made by these telescopes have not only advanced our scientific knowledge but also inspired countless individuals to look up at the night sky with a renewed sense of wonder.

The journey of space exploration is far from over. With each mission, we continue to push the boundaries of what we know and what we are capable of achieving. From the early days of launching satellites and sending the first humans into space to exploring distant planets with sophisticated robots and peering into the farthest reaches of the universe with advanced telescopes, our progress has been nothing short of extraordinary.

This book tells the story of our relentless quest to explore, understand, and eventually inhabit the final frontier. It is a testament to human curiosity, determination, and the unyielding desire to reach beyond the known. Through the lens of these 15 missions, we see the incredible journey of space exploration and the remarkable discoveries that have shaped our understanding of the universe. Each mission, a step forward in technology and knowledge, highlights the ongoing adventure of humanity's voyage into the great unknown.

MISSION I:
SPUTNIK I:
THE FIRST ARTIFICIAL SATELLITE

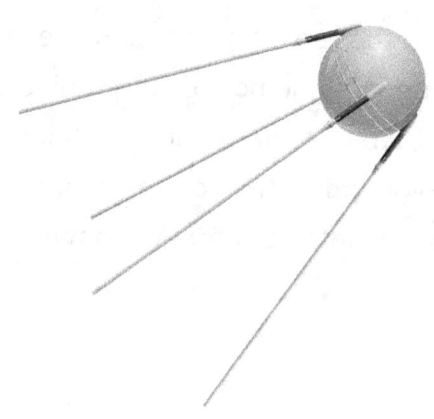

Introduction to Sputnik 1

On October 4, 1957, the world witnessed the dawn of a new era with the launch of Sputnik 1, the first artificial satellite to orbit the Earth. This historic event marked the beginning of the space age and set off a fierce competition between the United States and the Soviet Union, known as the Space Race. Sputnik 1 was launched by the Soviet Union, and its successful deployment into orbit had profound implications for both science and geopolitics.

The Launch of Sputnik 1:

Sputnik 1 was a relatively simple satellite, a spherical object with four external radio antennas designed to broadcast radio pulses. It weighed about 83.6 kilograms (184 pounds) and had a diameter of 58 centimeters (23 inches). Despite its simplicity, Sputnik 1's impact was anything but small. It orbited the Earth at a speed of approximately 29,000 kilometers per hour (18,000 miles per hour) and transmitted radio signals back to Earth, which could be received by radio operators around the globe. This constant beeping signal, detectable by anyone with a radio receiver, made it a dramatic and tangible demonstration of the Soviet Union's technological prowess.

The satellite's launch vehicle, the R-7 Semyorka, was a modified intercontinental ballistic missile (ICBM). The success of this rocket not only proved the feasibility of launching objects into space but also underscored the potential military applications of such technology. The public and political reactions around the world were immediate and profound, with many realizing the far-reaching implications of this achievement.

Global Reactions and Implications:

The successful launch of Sputnik 1 demonstrated the Soviet Union's advanced technological capabilities and served as a stark wake-up call to the United States, which had been working on its own satellite program. This event intensified the Cold War tensions, as it underscored the strategic advantages that space technology could offer, including potential surveillance and reconnaissance capabilities.

The global reaction was mixed. In the Soviet Union, Sputnik 1 was celebrated as a monumental achievement, bolstering national pride and solidifying the USSR's position as a leader in space exploration. In the United States, however, the reaction was one of urgency and concern. The American public and government officials feared that the Soviet Union had achieved a significant technological and strategic advantage. This fear was not just about space exploration but also about the implications for missile technology and military power.

U.S. Response and Space Race Intensification:

The United States responded swiftly to the launch of Sputnik 1. Recognizing the need to catch up, the U.S. government made significant changes to its policies and priorities. In 1958, President Dwight D. Eisenhower established the National Aeronautics and Space Administration (NASA), centralizing and accelerating the nation's space efforts. Additionally, the National Defense Education Act was passed, leading to increased funding for science and engineering education to ensure a future generation capable of advancing space technology.

This period marked the beginning of an intensified space race, with both superpowers striving to outdo each other with increasingly ambitious space missions. The launch of Sputnik 1 thus set off a series of rapid developments in space technology and exploration, fundamentally shaping the course of the Cold War and global scientific progress.

The Legacy of Sputnik 1:

Sputnik 1's journey ended when it re-entered Earth's atmosphere and burned up on January 4, 1958, after 22 days in orbit. However, its legacy lives on. The launch of Sputnik 1 not only marked the start of human space exploration but also paved the way for future satellite technology, which today plays a crucial role in communications, weather forecasting, navigation, and scientific research.

The success of Sputnik 1 demonstrated that space exploration was not just a possibility but a reality, one that could yield immense scientific and technological benefits. It set a precedent for international space collaboration and competition, driving advancements that continue to impact our world profoundly. The legacy of Sputnik 1 is a testament to the power of human ingenuity and the relentless pursuit of knowledge and exploration.

MISSION 2:
EXPLORER I:
THE FIRST SUCCESSFUL U.S. SATELLITE

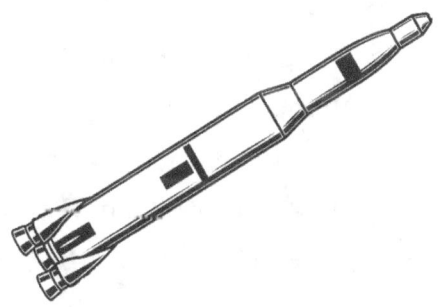

Introduction to Explorer 1:

Following the Soviet Union's launch of Sputnik 1, the United States entered the space race with its own milestone achievement. On January 31, 1958, the United States successfully launched Explorer 1, its first artificial satellite, into space. This mission not only marked the United States' entry into space exploration but also led to significant scientific discoveries that expanded our understanding of Earth's immediate space environment.

Development and Planning:

Explorer 1 was developed by a team of scientists led by Dr. James Van Allen at the University of Iowa and engineers at the Jet Propulsion Laboratory (JPL) under the direction of Dr. William Pickering. The project was part of the U.S. Army's response to the Soviet Union's successful Sputnik launches. The satellite was built as part of the Army's Jupiter-C rocket program, and its rapid development was a direct reaction to the Sputnik challenge.

The satellite was cylindrical, measuring about 2.03 meters (6 feet 7 inches) in length and 15.9

centimeters (6.25 inches) in diameter, and weighed approximately 14 kilograms (30.8 pounds). Explorer 1 was equipped with scientific instruments, including a cosmic ray detector designed by Dr. Van Allen, which would later contribute to the discovery of the Van Allen radiation belts.

The Launch of Explorer 1:

On January 31, 1958, at 10:48 p.m. EST, Explorer 1 was launched from Cape Canaveral, Florida, atop a Juno I rocket, a modified Redstone missile. The launch was a critical test of American capabilities and represented a significant step forward in the space race. As the rocket ascended and the satellite separated successfully from the final stage, it entered an elliptical orbit around Earth, reaching an apogee of about 2,550 kilometers (1,590 miles) and a perigee of about 360 kilometers (220 miles).

The successful launch of Explorer 1 marked a major triumph for the United States, providing a much-needed morale boost and proving that the nation could compete in space exploration. The satellite's instruments began transmitting data back to Earth, offering new insights into the space environment.

Scientific Discoveries and Contributions:

One of the most significant discoveries made by Explorer 1 was the identification of the Van Allen radiation belts. These belts are zones of charged particles trapped by Earth's magnetic field. Dr. Van Allen's cosmic ray detector showed a significant increase in radiation levels at certain altitudes, leading to the understanding that these belts are regions of high-energy particles, primarily protons and electrons, which pose a hazard to both satellites and astronauts.

This discovery was groundbreaking for several reasons. First, it provided critical information about the radiation environment that spacecraft and their occupants would encounter, informing the design of future missions and the development of protective measures. Second, it enhanced our understanding of Earth's magnetosphere and its interactions with solar wind and cosmic radiation.

Explorer 1 also carried a micrometeorite detector, which helped gather data on the presence and frequency of tiny particles in space. This information was crucial for assessing the risks posed by micrometeorite impacts on spacecraft, contributing to the design of more resilient satellites and space vehicles.

Soviet Response and Space Race Intensification:

The success of Explorer 1 had far-reaching implications for the Cold War and the space race. In the Soviet Union, the response to Explorer 1 was a mixture of concern and determination to maintain their lead in space exploration. The Soviet leadership recognized the importance of sustaining their technological superiority and responded by accelerating their own space initiatives.

In the wake of Explorer 1's success, the Soviet Union doubled down on its efforts to develop more advanced space technologies. This included planning for the launch of larger and more sophisticated satellites, as well as manned missions. The Soviet space program continued to push boundaries, leading to the successful launch of Luna 1 in January 1959, the first spacecraft to reach the vicinity of the Moon, and Luna 2 in September 1959, the first human-made object to impact the Moon.

The competition spurred by these achievements led to a series of rapid advancements from both superpowers. The Soviets, not wanting to be outdone, quickly followed with Vostok 1, which carried Yuri Gagarin, the first human in space, into orbit on April 12, 1961. This marked another significant milestone in the space race, demonstrating the Soviet Union's ability to send a human into space and safely return them to Earth.

This period of intense competition drove significant investments in research and development on both sides, leading to breakthroughs in rocket technology, spaceflight, and satellite communications. The space race became a central element of the Cold War, with each nation striving to demonstrate its technological and ideological superiority through space achievements.

The Legacy of Explorer 1:

Explorer 1 remained in orbit until March 31, 1970, long after its mission had ended. Over its 12-year orbit, it contributed valuable data that shaped our understanding of space and set the stage for future scientific endeavors. The Explorer program continued, with subsequent missions building on the success of the first satellite and exploring various aspects of space science and technology.

The legacy of Explorer 1 extends beyond its scientific contributions. It symbolized American ingenuity and determination, demonstrating that the United States could rise to the challenge posed by the Soviet Union and achieve significant milestones in space exploration. The mission's success also underscored the importance of international collaboration and the sharing of scientific knowledge, as researchers around the world benefited from the data and discoveries generated by the satellite.

Explorer 1's success reinforced the importance of scientific research in space exploration, highlighting the need for robust scientific instruments and rigorous experimentation. The mission's achievements paved the way for future scientific satellites and probes, many of which continue to explore our solar system and beyond, expanding our knowledge and inspiring new generations of scientists and engineers.

MISSION 3:

VOSTOK I:

THE FIRST HUMAN SPACEFLIGHT

Introduction to Vostok 1:

On April 12, 1961, the Soviet Union achieved a monumental milestone in human history by launching Vostok 1, the first manned spaceflight. Piloted by cosmonaut Yuri Gagarin, this mission marked the first time a human journeyed into outer space and orbited the Earth. Vostok 1 was a defining moment in the space race, showcasing Soviet space capabilities and setting the stage for future manned space exploration.

Development and Planning:

The development of Vostok 1 was part of the broader Soviet Vostok program, which aimed to send humans into space and safely return them to Earth. The program was spearheaded by Sergei Korolev, the chief designer and driving force behind the Soviet space program. The Vostok spacecraft was designed to accommodate a single cosmonaut and included life support systems to sustain them during the flight.

The Vostok spacecraft featured a spherical descent module and an instrument module. The spherical design was chosen for its simplicity and ability to withstand re-entry forces without needing precise orientation. The spacecraft also included a heat shield to protect it from the intense heat generated during re-entry into Earth's atmosphere. The cosmonaut's seat was equipped with an ejection mechanism to ensure a safe landing, as the descent module did not have a soft-landing capability.

The Launch of Vostok 1:

On the morning of April 12, 1961, Yuri Gagarin boarded Vostok 1 at the Baikonur Cosmodrome in Kazakhstan. At 9:07 a.m. Moscow time, the Vostok-K rocket lifted off, carrying Gagarin into space. As the spacecraft ascended, Gagarin famously exclaimed, "Poyekhali!" ("Let's go!"), capturing the excitement and significance of the moment.

Vostok 1 reached an altitude of approximately 327 kilometers (203 miles) and completed a single orbit around the Earth, traveling at a speed of around 27,400 kilometers per hour (17,000 miles per hour). The entire mission lasted approximately 108 minutes from launch to landing. During the flight, Gagarin reported feeling well and experienced weightlessness for the first time in human history. His calm demeanor and professionalism during the mission were instrumental in its success.

Scientific Discoveries and Contributions:

While the primary goal of Vostok 1 was to achieve human spaceflight, the mission also provided valuable scientific data. The flight demonstrated that humans could survive and function in the microgravity environment of space. Gagarin's physiological responses, including heart rate, blood pressure, and respiration, were monitored throughout the mission, providing essential information for future manned spaceflights.

Vostok 1 also tested the spacecraft's systems and re-entry procedures. The successful re-entry and recovery of the descent module validated the design and engineering of the Vostok spacecraft, paving the way for subsequent manned missions. Additionally, the mission highlighted the importance of human factors in spaceflight, such as the effects of isolation, confinement, and the psychological challenges of being in space.

U.S. Response and Space Race Intensification:

The success of Vostok 1 had a profound impact on the United States and the world. Yuri Gagarin's achievement was celebrated as a triumph of Sovietscience and technology, and he became an international hero. The mission solidified the Soviet

Union's leadership in the space race and demonstrated its ability to achieve significant milestones in space exploration.

The United States, already galvanized by the launch of Sputnik 1 and Explorer 1, intensified its efforts to catch up with the Soviet Union in the space race. President John F. Kennedy, recognizing the need for a bold response, delivered his famous speech on May 25, 1961, declaring the goal of landing a man on the Moon and returning him safely to Earth before the end of the decade. This announcement set the stage for the Apollo program and further escalated the space race between the two superpowers.

The American public and government officials were both inspired and alarmed by Gagarin's successful mission. It underscored the urgency of the United States developing its own manned spaceflight capabilities. NASA accelerated its efforts, pushing forward with Project Mercury, which aimed to send an American astronaut into space and bring them back safely.

The success of Vostok 1 also prompted the United States to invest more heavily in science and technology education, as well as in the development of more advanced rockets and spacecraft. This intensified focus on space exploration led to the rapid progression of U.S. space programs, culminating in the Mercury-Redstone 3 mission, which launched Alan Shepard into space on May 5, 1961, making him the first American in space.

The Legacy of Vostok 1:

Vostok 1's legacy is profound and far-reaching. Yuri Gagarin's historic flight not only marked the beginning of human space exploration but also inspired generations of scientists, engineers, and explorers. The mission demonstrated the feasibility of manned spaceflight and provided valuable data that informed the design of future spacecraft and missions.

Gagarin's flight also had significant cultural and political implications. He became a symbol of Soviet achievement and a global ambassador for space exploration. His flight fostered a sense of unity and excitement about the possibilities of space exploration and the potential for human achievement.

The success of Vostok 1 paved the way for subsequent manned space missions, including the Vostok program's continued exploration and the eventual development of the Soyuz spacecraft, which remains in use today. The mission's legacy is a testament to human curiosity, determination, and the unyielding desire to explore the unknown.

In conclusion, Vostok 1 was a defining moment in the history of space exploration. It marked the first human journey into space, demonstrated the feasibility of manned spaceflight, and inspired a global pursuit of scientific and technological advancements. Yuri Gagarin's courageous flight continues to be celebrated as a landmark achievement, and its legacy endures in the ongoing exploration of the cosmos.

MISSION 4:

VOSTOK 6:

THE FIRST WOMAN IN SPACE

Introduction to Vostok 6:

On June 16, 1963, the Soviet Union achieved another significant milestone in space exploration by launching Vostok 6, which carried Valentina Tereshkova, the first woman to fly in space. This historic mission not only showcased the Soviet Union's commitment to gender equality in space exploration but also marked a pivotal moment in the ongoing space race. Tereshkova's journey symbolized a major step forward in human spaceflight and the role of women in science and technology.

Development and Planning:

The Vostok program, under the direction of Sergei Korolev, aimed to achieve various spaceflight milestones. After the success of Vostok 1, which carried Yuri Gagarin, the first human in space, the program sought to send the first woman into orbit. Valentina Tereshkova, an amateur parachutist and textile worker, was selected from over 400 applicants and five finalists for her skills and background.

The Vostok 6 spacecraft was similar to its predecessors, designed to accommodate a single cosmonaut. It featured a spherical descent module for re-entry and an instrument module that housed life support systems, communication equipment, and scientific instruments. The mission was meticulously planned, with extensive training for Tereshkova, including simulations and physical conditioning, to ensure she was prepared for the challenges of spaceflight.

The Launch of Vostok 6:

On the morning of June 16, 1963, Valentina Tereshkova boarded the Vostok 6 spacecraft at the Baikonur Cosmodrome in Kazakhstan. At 9:29 a.m. Moscow time, the Vostok-K rocket lifted off, carrying Tereshkova into space. During the ascent, she famously communicated, "Yuri, tell me, is everything ready?" to which she received an affirmative response, acknowledging the path paved by Yuri Gagarin.

Vostok 6 successfully entered orbit, reaching an altitude of 230 kilometers (143 miles). Tereshkova orbited the Earth 48 times over nearly three days, spending a total of 71 hours in space. Her call sign for the mission was "Chaika," meaning "Seagull" in Russian. During the mission, she conducted various tests, maintained a flight log, and took photographs of the Earth, which later contributed to understanding the atmosphere.

Scientific Discoveries and Contributions:

Vostok 6 provided valuable data on the effects of spaceflight on the female body. Tereshkova's physiological responses, such as heart rate, blood pressure, and respiration, were monitored throughout the mission. Her experiences contributed to the broader understanding of how spaceflight impacts human health and provided insights for future missions involving female astronauts.

The mission also demonstrated the effectiveness of the Vostok spacecraft's design and systems, further validating the engineering and technological advancements achieved by the Soviet space program. Tereshkova's successful re-entry and landing in Kazakhstan, where she was greeted as a national hero, underscored the mission's success.

Advancing Gender Equality:

The success of Vostok 6 was a significant achievement for the Soviet Union and an important step in advancing gender equality in space exploration. Valentina Tereshkova's flight demonstrated that women could effectively participate in and contribute to space missions, paving the way for future generations of female astronauts. Her journey symbolized the breaking of gender barriers in one of the most challenging environments known to humanity.

Tereshkova's achievement highlighted the importance of diversity and inclusion in advancing scientific knowledge and technological innovation. Her flight remains a symbol of human progress and the ongoing quest to explore the unknown. The inclusion of women in space programs became an increasingly important issue for NASA and other space agencies worldwide, leading to more opportunities for women in STEM fields.

The Legacy of Vostok 6:

Valentina Tereshkova's historic flight on Vostok 6 left an enduring legacy. As the first woman in space, she became a global icon and an inspiration for women in science, technology, engineering, and mathematics (STEM) fields. Her achievement demonstrated the potential for women to contribute significantly to space exploration and broke gender barriers in one of the most challenging environments known to humanity.

The success of Vostok 6 paved the way for the inclusion of women in space programs around the world. It highlighted the importance of diversity and inclusion in advancing scientific knowledge and technological innovation. Tereshkova's flight remains a symbol of human progress and the ongoing quest to explore the unknown.

In conclusion, Vostok 6 was a landmark achievement in the history of space exploration. It marked the first time a woman traveled to space, demonstrated the capabilities of the Vostok spacecraft, and provided valuable data for future missions. Valentina Tereshkova's courageous flight continues to be celebrated as a monumental step forward in human spaceflight, reminding us of the limitless possibilities of space exploration and the boundless potential of human ingenuity.

MISSION 5:
APOLLO 8:
THE FIRST MANNED MISSION TO ORBIT THE MOON

Introduction to Apollo 8:

On December 21, 1968, NASA launched Apollo 8, the first manned spacecraft to orbit the Moon. This mission was a significant milestone in the Apollo program and marked a pivotal moment in human space exploration. Apollo 8 was not only the first manned mission to leave Earth's orbit but also the first to travel to the Moon and return safely. The success of Apollo 8 paved the way for future lunar landings, including the historic Apollo 11 mission.

Development and Planning:

Apollo 8 was originally intended to be an Earth-orbit mission to test the Lunar Module, but delays in its development led to a change in plans. Instead, NASA decided to send Apollo 8 on a bold mission to orbit the Moon. This decision was influenced by the desire to achieve a major milestone before the end of 1968, as well as to demonstrate the capabilities of the Saturn V rocket and the Apollo spacecraft in deep space.

The mission was commanded by Frank Borman, with James Lovell as the Command Module Pilot and William Anders as the Lunar Module Pilot. The crew underwent extensive training to prepare for the mission, including simulations of the lunar orbit and re-entry procedures. The Apollo spacecraft consisted of the Command Module, named "Charlie Brown," and the Service Module, which provided propulsion, electrical power, and life support systems.

The Launch of Apollo 8:

Apollo 8 launched from Kennedy Space Center in Florida on December 21, 1968, atop a Saturn V rocket. The Saturn V was the most powerful rocket ever built, standing 110 meters (363 feet) tall and capable of generating 7.6 million pounds of thrust. The launch was flawless, and the spacecraft entered a parking orbit around Earth before initiating the Trans-Lunar Injection (TLI) burn to set it on a trajectory toward the Moon.

The TLI burn, performed by the third stage of the Saturn V, accelerated Apollo 8 to a velocity of approximately 39,400 kilometers per hour (24,500 miles per hour), allowing it to break free of Earth's gravitational pull. The spacecraft then coasted for three days toward the Moon, during which time the crew conducted various system checks and navigation adjustments.

Lunar Orbit and Historic Achievements:

On December 24, 1968, Apollo 8 entered lunar orbit, becoming the first manned spacecraft to do so. The crew conducted ten orbits of the Moon over the course of 20 hours, capturing stunning photographs of the lunar surface and the iconic "Earthrise" image, which showed Earth rising above the lunar horizon. This image became one of the most famous photographs in history, symbolizing the fragility and unity of our planet.

During their time in lunar orbit, the crew conducted a detailed survey of potential landing sites for future missions. They provided valuable information about the lunar terrain, which was crucial for the planning of the Apollo 11 landing. The mission also demonstrated the feasibility of navigating to and from the Moon, validating the techniques and procedures that would be used in subsequent lunar missions.

U.S. Response and Space Race Intensification:

The success of Apollo 8 had a profound impact on the United States and the world. It marked a significant victory for NASA and the United States in the space race against the Soviet Union. The mission showcased American technological prowess and served as a testament to the dedication and ingenuity of the engineers, scientists, and astronauts involved in the Apollo program.

The American public and government officials were thrilled by the achievement, which provided a much-needed boost to national morale during a tumultuous period in U.S. history. The success of Apollo 8 reinforced the commitment to achieving President Kennedy's goal of landing a man on the Moon and returning him safely to Earth before the end of the decade. It also demonstrated the effectiveness of the Saturn V rocket and the Apollo spacecraft, instilling confidence in the feasibility of a lunar landing.

Soviet Response and Space Race Intensification:

The success of Apollo 8 did not go unnoticed by the Soviet Union. The Soviets, who had achieved numerous firsts in space exploration, including launching the first artificial satellite and sending the first human into space, now faced a significant challenge. The Apollo 8 mission demonstrated that the United States had made substantial progress in its space program, and the Soviets needed to respond to maintain their leadership position.

The Soviet space program intensified its efforts to advance its own lunar missions. Despite their initial achievements, the Soviets had encountered setbacks with their N1 rocket, which was intended to compete with the Saturn V for manned lunar missions. The N1 rocket faced a series of failures, delaying the Soviet lunar program and preventing it from matching the Apollo program's accomplishments.

In response to Apollo 8, the Soviets focused on enhancing their existing space capabilities. They accelerated the development of their Soyuz spacecraft, which would become a cornerstone of their manned space missions. The Soviet Union also continued to focus on lunar exploration through unmanned missions, such as the Luna program, which included lunar orbiters, landers, and rovers.

The success of Apollo 8 further highlighted the competitive nature of the space race, driving both superpowers to push the boundaries of space exploration. The Soviet Union's response underscored the importance of maintaining technological advancements and showcased their determination to continue contributing to space exploration despite the setbacks.

The Legacy of Apollo 8:

Apollo 8's legacy is immense, both in terms of its scientific and cultural impact. The mission demonstrated the feasibility of human spaceflight to the Moon and provided valuable data for future lunar missions. The "Earthrise" photograph captured during the mission remains one of the most iconic images of the 20th century, inspiring a greater awareness of Earth's environment and the importance of protecting our planet.

The success of Apollo 8 also paved the way for the subsequent Apollo missions, culminating in the historic Apollo 11 Moon landing in July 1969. The mission's achievements highlighted the potential of human space exploration and set the stage for further exploration of the Moon, Mars, and beyond.

In conclusion, Apollo 8 was a landmark achievement in the history of space exploration. It marked the first time humans traveled to and orbited the Moon, demonstrated the capabilities of the Apollo spacecraft and Saturn V rocket, and provided critical data for future lunar missions. The mission's success inspired a sense of unity and wonder, reminding us of the limitless possibilities of human ingenuity and the boundless potential of space exploration.

MISSION 6:

APOLLO II:

THE FIRST MANNED MOON LANDING

Introduction to Apollo 11

On July 20, 1969, the world watched in awe as Apollo 11 became the first mission to successfully land humans on the Moon. This historic achievement marked a monumental milestone in space exploration and fulfilled President John F. Kennedy's 1961 goal of landing a man on the Moon and returning him safely to Earth before the decade's end. The Apollo 11 mission, commanded by Neil Armstrong with astronauts Edwin "Buzz" Aldrin and Michael Collins, symbolized the pinnacle of American technological prowess and human determination.

Development and Planning:

The Apollo program was initiated by NASA to achieve the goal of manned lunar landing. Following the successes of earlier missions, such as Apollo 8, which orbited the Moon, and Apollo 10, which performed a "dress rehearsal" for the landing, Apollo 11 was tasked with the actual landing. The mission required extensive planning, engineering, and testing to ensure the safety and success of the astronauts.

The Apollo 11 spacecraft consisted of three modules: the Command Module (Columbia), the Service Module, and the Lunar Module (Eagle). The

Command Module housed the crew during launch, lunar orbit, and re-entry. The Service Module provided propulsion, electrical power, and storage for various consumables. The Lunar Module was designed to land on the Moon's surface and return to lunar orbit.

The astronauts underwent rigorous training to prepare for the mission, including simulations of the lunar landing and extravehicular activities (EVA). The mission's flight plan was meticulously crafted to address every possible contingency, ensuring that the crew could handle any challenges that arose during their journey.

The Launch of Apollo 11:

Apollo 11 launched from Kennedy Space Center in Florida on July 16, 1969, atop a Saturn V rocket. The Saturn V, the most powerful rocket ever built, was crucial in propelling the spacecraft toward the Moon. The launch was flawless, and the spacecraft entered Earth orbit before initiating the Trans-Lunar Injection (TLI) burn, which set it on course for the Moon.

The TLI burn, performed by the third stage of the Saturn V, accelerated Apollo 11 to a velocity of approximately 39,000 kilometers per hour (24,000 miles per hour), allowing it to break free of Earth's gravitational pull. The spacecraft then coasted for three days toward the Moon, during which time the crew conducted various system checks and prepared for the lunar landing.

Lunar Landing and Historic Achievements:

On July 20, 1969, the Lunar Module, Eagle, separated from the Command Module, Columbia, and began its descent to the Moon's surface. As the Eagle descended, Neil Armstrong and Buzz Aldrin encountered a few challenges, including a navigational error that took them off course and a fuel shortage that added urgency to their landing. Despite these challenges, Armstrong manually piloted the Eagle to a safe landing site in the Sea of Tranquility.

At 20:17 UTC, Neil Armstrong radioed back to Mission Control, "Houston, Tranquility Base here. The Eagle has landed." This moment marked the successful landing of the first humans on the Moon. A few hours later, on July 21, 1969, Armstrong descended the ladder and set foot on the lunar surface, uttering the famous words, "That's one small step for man, one giant leap for mankind." Buzz Aldrin soon followed, and together they spent approximately two and a half hours exploring the lunar surface, collecting samples, and conducting experiments.

U.S. Triumph and Global Reaction:

The success of Apollo 11 was a triumph for the United States and a defining moment in the space race against the Soviet Union. It showcased the nation's technological capabilities and affirmed its leadership in space exploration. The mission's success provided a significant morale boost to the American public and solidified NASA's reputation as a world-leading space agency.

The American public and government officials celebrated the achievement with great enthusiasm. Parades, speeches, and public events were held to honor the astronauts and the thousands of engineers, scientists, and support staff who made the mission possible. The successful landing and safe return of Apollo 11 fulfilled President Kennedy's vision and demonstrated the power of human ingenuity and perseverance.

Soviet Response and Space Race Intensification:

The success of Apollo 11 did not go unnoticed by the Soviet Union. The Soviets, who had achieved numerous firsts in space exploration, including launching the first artificial satellite and sending the first human into space, now faced a significant challenge. The Apollo 11 mission demonstrated that the United States had made substantial progress in its space program, and the Soviets needed to respond to maintain their leadership position.

In response, the Soviet Union intensified its efforts to advance its own space program. Despite setbacks with their N1 rocket, which was intended for manned lunar missions, the Soviets continued to focus on their space achievements. They pursued lunar exploration through unmanned missions, such as the Luna program, which included lunar orbiters, landers, and rovers.

Some historians argue that the success of Apollo 11 and the subsequent lunar landings played a role in the eventual collapse of the Soviet Union by highlighting the technological and economic disparities between the two superpowers. The significant investment required for the space race placed a strain on the Soviet economy, contributing to broader systemic issues that eventually led to the dissolution of the Soviet Union in 1991.

The Soviets also concentrated on enhancing their existing space capabilities. They accelerated the development of their Soyuz spacecraft, which became a cornerstone of their manned space missions. The Soviet space program remained a significant force in space exploration, contributing to scientific knowledge and technological advancements.

The Legacy of Apollo 11:

Apollo 11's legacy is profound and far-reaching. The mission demonstrated the feasibility of human spaceflight to the Moon and provided valuable data for future lunar missions. The iconic image of Neil Armstrong's footprint on the lunar surface and the "Buzz Aldrin on the Moon" photograph captured during the mission remain some of the most recognizable symbols of human achievement.

The success of Apollo 11 also paved the way for subsequent Apollo missions, including the exploration of different lunar regions and the collection of extensive scientific data. The mission's achievements highlighted the potential of human space exploration and set the stage for further exploration of the Moon, Mars, and beyond.

In conclusion, Apollo 11 was a landmark achievement in the history of space exploration. It marked the first time humans traveled to and landed on the Moon, demonstrated the capabilities of the Apollo spacecraft and Saturn V rocket, and provided critical data for future lunar missions. The mission's success inspired a sense of unity and wonder, reminding us of the limitless possibilities of human ingenuity and the boundless potential of space exploration. As Neil Armstrong famously said, "That's one small step for man, one giant leap for mankind."

MISSION 7:
VIKING I:

THE FIRST SUCCESSFUL MISSION TO LAND ON MARS

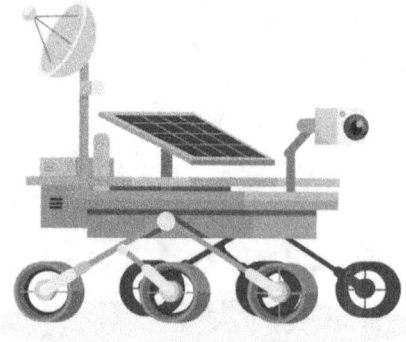

Introduction to Viking 1:

On July 20, 1976, NASA's Viking 1 became the first spacecraft to successfully land on Mars and transmit high-resolution images of the Martian surface back to Earth. The Viking 1 mission, part of the broader Viking program, marked a significant milestone in planetary exploration. It provided the first detailed images of Mars, conducted a variety of scientific experiments, and laid the groundwork for future missions to the Red Planet.

Development and Planning:

The Viking program was initiated by NASA in the early 1970s with the goal of sending orbiters and landers to Mars to search for signs of life, study the planet's surface and atmosphere, and provide data for future missions. The Viking 1 spacecraft consisted of two main components: an orbiter and a lander. The orbiter was designed to map the Martian surface and relay communications between the lander and Earth, while the lander was equipped with scientific instruments to conduct experiments directly on the Martian surface.

The Viking landers were designed to operate in the harsh Martian environment, with instruments to measure the atmospheric composition, temperature, and pressure. The lander also carried a suite of scientific experiments, including a gas chromatograph-mass spectrometer to analyze soil samples, a biological laboratory to search for signs of life, and cameras to capture images of the Martian landscape.

The Launch of Viking 1:

Viking 1 was launched from Kennedy Space Center in Florida on August 20, 1975, aboard a Titan IIIE rocket. The spacecraft embarked on a nearly year-long journey to Mars, arriving in orbit around the planet on June 19, 1976. After spending several weeks in orbit to select a suitable landing site, Viking 1's lander separated from the orbiter and descended to the surface of Mars.

The landing site chosen for Viking 1 was Chryse Planitia, a relatively flat and smooth region in the northern hemisphere of Mars. On July 20, 1976, the lander successfully touched down on the Martian surface, becoming the first spacecraft to land on Mars and transmit data back to Earth. This historic event occurred exactly seven years after the Apollo 11 Moon landing, underscoring NASA's continued commitment to space exploration.

Scientific Discoveries and Contributions:

Viking 1's scientific experiments provided a wealth of information about Mars. The lander's cameras captured the first high-resolution images of the Martian surface, revealing a barren, rocky landscape with scattered boulders and a thin, reddish atmosphere. These images helped scientists understand the geology and morphology of Mars and provided critical data for future missions.

One of the primary goals of the Viking program was to search for signs of life on Mars. The lander conducted several biological experiments designed to detect microbial life in the Martian soil. These experiments included the gas exchange experiment, the labeled release experiment, and the pyrolytic release experiment. While the results of these experiments were inconclusive and remain a topic of debate, they provided valuable insights into the chemical composition and reactivity of Martian soil.

Viking 1 also conducted atmospheric measurements, revealing that the Martian atmosphere is composed primarily of carbon dioxide with traces of nitrogen and argon. The lander measured surface temperatures ranging from -120°C to -14°C (-184°F to 7°F) and detected seasonal changes in atmospheric pressure, suggesting the presence of polar ice caps that grow and shrink with the changing seasons.

U.S. Triumph and Global Reaction:

The success of Viking 1 was a major triumph for NASA and the United States. The mission demonstrated America's ability to achieve complex planetary exploration goals and provided a significant boost to national pride and scientific interest in Mars. The data and images returned by Viking 1 captivated the public and scientists alike, inspiring a renewed interest in the Red Planet.

The American public and scientific community celebrated the achievement, recognizing the mission's importance in advancing our understanding of Mars and laying the groundwork for future exploration. Viking 1's success reinforced NASA's reputation as a world leader in space exploration and showcased the agency's ability to conduct ambitious and scientifically valuable missions.

Soviet Response and Space Race Intensification:

The success of Viking 1 was closely monitored by the Soviet Union, which had its own ambitious plans for Mars exploration. The Soviet space program had achieved significant milestones in lunar and planetary exploration, including the first successful mission to Venus with Venera 7 in 1970. However, the Soviets had encountered several challenges with their Mars missions, including the failures of Mars 4 and Mars 7 in 1973.

The achievements of Viking 1 underscored the technological advancements and scientific capabilities of NASA and prompted the Soviet Union to redouble its efforts in planetary exploration. Some historians argue that the success of U.S. space missions, including Apollo 11 and Viking 1, highlighted the technological and economic disparities between the two superpowers, contributing to the eventual collapse of the Soviet Union.

Despite these challenges, the Soviet Union continued to pursue Mars exploration, developing new missions such as the Phobos program in the late 1980s. These missions aimed to explore the Martian moons Phobos and Deimos and gather more data about Mars itself. The Soviet space program remained a significant force in space exploration, contributing valuable scientific knowledge and technological advancements.

The Legacy of Viking 1:

Viking 1's legacy is profound and enduring. The mission provided the first detailed images and data from the surface of Mars, greatly enhancing our understanding of the planet's geology, atmosphere, and potential for life. The data collected by Viking 1 laid the groundwork for future Mars missions, including the Mars Pathfinder, Spirit and Opportunity rovers, and the Curiosity and Perseverance rovers.

The success of Viking 1 also demonstrated the feasibility and value of robotic planetary exploration. The mission's achievements highlighted the potential of unmanned spacecraft to conduct complex scientific investigations and gather data from distant worlds. Viking 1's legacy continues to inspire new generations of scientists, engineers, and explorers, driving the ongoing quest to explore Mars and beyond.

In conclusion, Viking 1 was a landmark achievement in the history of planetary exploration. It marked the first successful mission to land on Mars and provided invaluable data about the Martian surface and atmosphere. The mission's success inspired a sense of wonder and curiosity about Mars, reminding us of the limitless possibilities of space exploration and the boundless potential of human ingenuity.

MISSION 8:
VOYAGER I AND 2:
OUR FIRST JOURNEY TO INTERSTELLAR SPACE

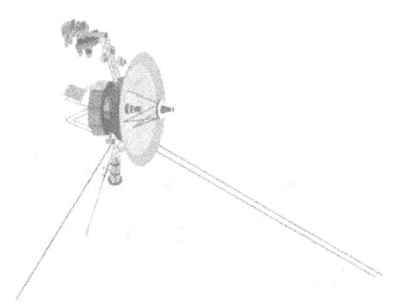

Introduction to Voyager 1 and 2:

Launched by NASA in 1977, the twin Voyager spacecraft, Voyager 1 and Voyager 2, embarked on a grand tour of the outer planets, providing humanity with its first detailed images and data from Jupiter, Saturn, Uranus, and Neptune. These missions expanded our understanding of the solar system and beyond, with Voyager 1 becoming the first human-made object to enter interstellar space in 2012. The Voyager missions remain some of the most ambitious and successful space exploration endeavors in history.

Development and Planning:

The Voyager program was conceived during the early 1970s as an ambitious plan to take advantage of a rare planetary alignment that occurs approximately every 176 years. This alignment would allow a spacecraft to use gravitational assists to travel from one planet to the next without needing excessive amounts of propulsion. The primary mission was to explore Jupiter and Saturn, with extended missions to Uranus and Neptune for Voyager 2.

Each Voyager spacecraft was equipped with a suite of scientific instruments designed to study the atmospheric conditions, magnetic fields, rings, and moons of the planets they encountered. Instruments included imaging systems, magnetometers, plasma detectors, and various spectrometers to analyze the chemical compositions and physical properties of planetary atmospheres and surfaces.

The Launch and Journey:

Voyager 2 was launched first on August 20, 1977, followed by Voyager 1 on September 5, 1977. Although Voyager 1 was launched later, it was placed on a faster trajectory, allowing it to reach Jupiter and Saturn before its twin. The spacecraft were powered by radioisotope thermoelectric generators (RTGs), which provided a reliable source of electrical power essential for the long-duration mission.

Encounters with the Outer Planets:

Jupiter (1979):
Both spacecraft conducted detailed studies of Jupiter's atmosphere, magnetic field, and moons. The Voyagers revealed the complexity of Jupiter's clouds and storms, including the Great Red Spot, and

discovered volcanic activity on the moon Io, which was the first active volcano seen on another body in the solar system.

Saturn (1980-1981):
The Voyager missions provided unprecedented data on Saturn's rings, atmosphere, and moons. Voyager 1 discovered intricate ring structures and found that Saturn's moon Titan had a thick, nitrogen-rich atmosphere. Voyager 2 continued the study of Saturn, providing detailed images and data.

Uranus (1986):
Voyager 2 is the only spacecraft to have visited Uranus. It discovered 11 new moons, studied the planet's unique tilted magnetic field, and provided insights into its atmosphere and rings. The spacecraft's observations of Uranus's moon Miranda revealed a surface with giant canyons and unusual geological features.

Neptune (1989):
Voyager 2's flyby of Neptune revealed the planet's dynamic atmosphere, including the Great Dark Spot, a massive storm comparable to Jupiter's Great Red Spot. The mission also discovered six new moons and provided detailed images of Neptune's rings and its largest moon, Triton, which was found to have geysers erupting nitrogen gas.

Interstellar Mission:

After completing their primary missions, both Voyagers continued their journey towards the edge of the solar system. Voyager 1 entered interstellar space on August 25, 2012, becoming the first human-made object to do so. It provided invaluable data on the boundary between the solar system and interstellar space, known as the heliopause. Voyager 2 followed, entering interstellar space on November 5, 2018.

Scientific Discoveries and Contributions:

The Voyager missions have contributed immensely to our understanding of the solar system. They provided detailed data on the outer planets and their moons, revealing complex atmospheres, magnetic fields, and geological activity. The missions also demonstrated the feasibility of using gravity assists to explore multiple planets with a single spacecraft. The data collected by the Voyagers has led to numerous scientific discoveries and has been instrumental in planning future missions to the outer planets and beyond. The missions have expanded our knowledge of planetary science, space physics, and the conditions at the edge of the solar system.

Current Status and Location:

As of 2024, Voyager 1 is approximately 15.144.685,480 miles (24.4 billion kilometers) from Earth, traveling through interstellar space. Voyager 2 is about 12.649.978,688 miles (20.4 billion kilometers) from Earth. Both spacecraft continue to send data back to NASA, contributing to our understanding of the environment beyond our solar system. You can check the current distance of the Voyager spacecraft by visiting the <u>NASA Voyager website</u> to see how much farther they have traveled since the writing of this book (<u>Voyager JPL NASA</u>) (<u>NASA Science</u>).

The Legacy of Voyager 1 and 2:

The legacy of the Voyager missions is enduring and far-reaching. They remain some of the most successful and iconic missions in the history of space exploration. The data collected by the Voyagers continues to be analyzed and has provided a foundation for future missions to the outer planets and beyond.

The Voyagers' journey into interstellar space represents a significant milestone in human exploration, symbolizing our quest to understand the cosmos and our place within it. Their ongoing mission continues to provide valuable scientific data and inspire a sense of wonder and curiosity about the universe.

In conclusion, the Voyager missions were landmark achievements in space exploration. They provided unprecedented data on the outer planets, expanded our understanding of the solar system, and demonstrated the potential of robotic spacecraft to conduct long-duration missions. The success of Voyager 1 and 2 continues to inspire and inform future generations of space explorers, reminding us of the limitless possibilities of space exploration and the boundless potential of human ingenuity.

MISSION 9:
HUBBLE SPACE TELESCOPE

REVOLUTIONIZING OUR UNDERSTANDING OF THE UNIVERSE

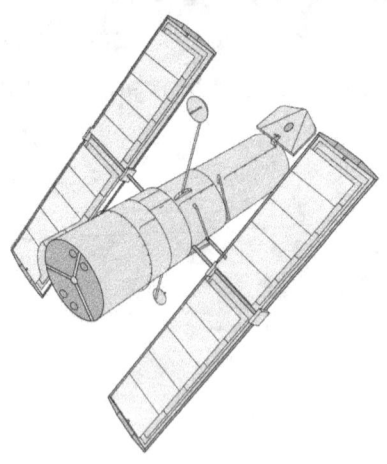

Introduction to the Hubble Space Telescope:

Launched on April 24, 1990, aboard the Space Shuttle Discovery, the Hubble Space Telescope (HST) has been one of the most influential instruments in the history of astronomy. Positioned above Earth's atmosphere, Hubble has provided unprecedented clarity and detail in its observations, significantly enhancing our understanding of the universe. Over its decades-long mission, Hubble has captured breathtaking images and made groundbreaking discoveries that have revolutionized various fields of astronomy and astrophysics.

Development and Planning:

The concept of a space-based telescope dates back to the 1940s, but it wasn't until the 1970s that serious planning for the Hubble Space Telescope began. NASA collaborated with the European Space Agency (ESA) to develop Hubble, with ESA providing the solar arrays and one of the primary scientific instruments, the Faint Object Camera.

Hubble's design includes a 2.4-meter primary mirror and five main instruments, which have been upgraded and replaced over several servicing missions conducted by Space Shuttle astronauts.

These instruments include cameras, spectrographs, and fine guidance sensors that have enabled Hubble to capture detailed images across multiple wavelengths of light, from ultraviolet to near-infrared.

The Launch and Initial Challenges:

Hubble was launched into low Earth orbit on April 24, 1990, and deployed the following day. However, soon after its deployment, scientists discovered a flaw in the primary mirror's shape, which caused spherical aberration and blurred the images. This flaw was due to a miscalibration during the mirror's manufacturing process.

In 1993, astronauts aboard the Space Shuttle Endeavour carried out the first servicing mission (STS-61) to install corrective optics, known as COSTAR (Corrective Optics Space Telescope Axial Replacement), and replace the Wide Field and Planetary Camera with an upgraded version. These corrections successfully restored Hubble's vision, allowing it to fulfill its scientific potential.

Scientific Discoveries and Contributions:

Hubble has made numerous significant contributions to our understanding of the universe. Some of its most notable achievements include:

- **Deep Field Images:** Hubble's deep field observations, such as the Hubble Deep Field (1995) and the Hubble Ultra Deep Field (2004), provided the deepest views of the universe ever obtained, revealing thousands of galaxies in a tiny patch of sky. These images have allowed astronomers to study galaxy formation and evolution over cosmic time.

- **Accelerating Universe:** Observations of distant supernovae by Hubble played a crucial role in the discovery that the expansion of the universe is accelerating, leading to the concept of dark energy, a mysterious force driving this acceleration.

- **Exoplanet Atmospheres:** Hubble has been instrumental in the study of exoplanets, including the detection of atmospheres around these distant worlds. It has identified the presence of water vapor, methane, and other molecules in exoplanet atmospheres, providing insights into their potential habitability.

- **Star Formation and Death:** Hubble's high-resolution images have allowed astronomers to study the life cycles of stars in unprecedented detail, from star formation in nebulae to the explosive deaths of stars as supernovae.

- **Deep Field Images:** Hubble's deep field observations, such as the Hubble Deep Field (1995) and the Hubble Ultra Deep Field (2004), provided the deepest views of the universe ever obtained, revealing thousands of galaxies in a tiny patch of sky. These images have allowed astronomers to study galaxy formation and evolution over cosmic time.

- **Accelerating Universe:** Observations of distant supernovae by Hubble played a crucial role in the discovery that the expansion of the universe is accelerating, leading to the concept of dark energy, a mysterious force driving this acceleration.

- **Exoplanet Atmospheres:** Hubble has been instrumental in the study of exoplanets, including the detection of atmospheres around these distant worlds. It has identified the presence of water vapor, methane, and other molecules in exoplanet atmospheres, providing insights into their potential habitability.

- **Star Formation and Death:** Hubble's high-resolution images have allowed astronomers to study the life cycles of stars in unprecedented detail, from star formation in nebulae to the explosive deaths of stars as supernovae.

Current Status and Future Prospects:

Hubble continues to operate and produce valuable scientific data, despite being well beyond its original mission lifetime. The last servicing mission, STS-125 in 2009, equipped Hubble with new instruments and repaired existing ones, extending its operational life. The telescope is expected to continue functioning well into the 2020s, working in tandem with the James Webb Space Telescope (JWST), which was successfully launched on December 25, 2021. The JWST will complement Hubble's observations by providing even deeper insights into the universe, particularly in the infrared spectrum.

The Legacy of the Hubble Space Telescope:

The legacy of the Hubble Space Telescope is immense and far-reaching. It has transformed our understanding of the universe, provided stunning visual images that have captured the public's imagination, and inspired generations of scientists and space enthusiasts. Hubble's discoveries have led to over 18,000 scientific papers, making it one of the most productive scientific instruments ever built.

Hubble's contributions to science and its iconic status in space exploration will be remembered long after it ceases operations. The data it has collected will continue to be analyzed, yielding new discoveries for years to come.

Personally, I believe Hubble is among the top three missions in the history of space exploration. Its profound impact on our understanding of the universe and its ability to capture the public's imagination through stunning imagery make it a cornerstone of modern astronomy.

In conclusion, the Hubble Space Telescope has revolutionized our understanding of the universe. Its detailed observations have provided unprecedented insights into the cosmos, from the formation of galaxies to the discovery of dark energy. The success and legacy of Hubble continue to inspire and inform future generations of astronomers and space explorers, demonstrating the boundless potential of human ingenuity and scientific curiosity.

MISSION 10:
CASSINI-HUYGENS

EXPLORING SATURN AND ITS MOONS IN UNPRECEDENTED DETAIL

Introduction to Cassini-Huygens:

Launched on October 15, 1997, the Cassini-Huygens mission was a collaborative project between NASA, the European Space Agency (ESA), and the Italian Space Agency (ASI). The mission's primary objectives were to study Saturn, its rings, and its diverse moons, providing humanity with a wealth of scientific data and stunning images. Cassini-Huygens revolutionized our understanding of the Saturnian system and made numerous groundbreaking discoveries during its 20-year mission.

Development and Planning:

The Cassini-Huygens mission was named after two prominent astronomers: Giovanni Cassini, who discovered several of Saturn's moons and the gap in Saturn's rings (known as the Cassini Division), and Christiaan Huygens, who discovered Titan, Saturn's largest moon. The mission consisted of two main components: the Cassini orbiter, built by NASA, and the Huygens probe, built by ESA.

The Cassini orbiter was equipped with a suite of 12 scientific instruments designed to study Saturn's atmosphere, rings, magnetic field, and moons. The

Huygens probe was designed to detach from the orbiter and descend through the atmosphere of Titan, Saturn's largest moon, to study its surface and atmospheric composition.

The Launch and Journey:

Cassini-Huygens was launched from Cape Canaveral, Florida, aboard a Titan IVB/Centaur rocket on October 15, 1997. The spacecraft followed a complex trajectory that included gravity assist flybys of Venus, Earth, and Jupiter to gain the necessary speed to reach Saturn. After a seven-year journey, Cassini entered orbit around Saturn on July 1, 2004.

Scientific Discoveries and Contributions:

Titan Exploration:

One of the most significant achievements of the Cassini-Huygens mission was the exploration of Titan. On January 14, 2005, the Huygens probe descended through Titan's thick atmosphere and landed on its surface, becoming the first human-made object to land on a moon in the outer solar system. The probe transmitted data and images for about 90 minutes, revealing a landscape with river channels, lakes, and dunes made of hydrocarbons.

Enceladus Plumes:

Cassini discovered geysers of water vapor and ice particles erupting from the south polar region of Enceladus, one of Saturn's moons. These plumes suggested the presence of a subsurface ocean and provided evidence of hydrothermal activity, making Enceladus a prime candidate in the search for extraterrestrial life.

Saturn's Rings:

Cassini provided unprecedented insights into the structure and dynamics of Saturn's rings. It captured detailed images of the rings' intricate structures, including propeller-like features caused by moonlets within the rings and the changing patterns of ring spokes. The mission also studied the composition and age of the rings, enhancing our understanding of their origins.

Seasonal Changes:

Cassini observed seasonal changes on Saturn and its moons over its 13-year mission in orbit. It captured the changing colors of Saturn's atmosphere, the formation and dissipation of storms, and the effects of seasonal variations on Titan's methane lakes and seas.

The Grand Finale:

In its final phase, known as the Grand Finale, Cassini performed a series of daring orbits that took it between Saturn and its innermost ring. These maneuvers provided unique data on Saturn's gravity, magnetic field, and the composition of its rings. On September 15, 2017, Cassini made a controlled descent into Saturn's atmosphere, transmitting data until it was destroyed by the intense heat and pressure.

Current Status and Legacy:

Although the Cassini spacecraft is no longer operational, its legacy lives on through the vast amount of data it collected, which continues to be analyzed by scientists worldwide. The mission's discoveries have provided valuable insights into the Saturnian system, planetary formation, and the potential for life beyond Earth.

The Legacy of Cassini-Huygens:

The Cassini-Huygens mission significantly advanced our knowledge of Saturn and its moons, revealing the complexity and diversity of the Saturnian system. It showcased the power of international collaboration in space exploration and set new standards for planetary science missions. The stunning images and groundbreaking discoveries from Cassini-Huygens continue to inspire scientists and the public, highlighting the importance of exploration and the quest to understand our place in the universe.

In conclusion, the Cassini-Huygens mission was a landmark achievement in space exploration. It provided unprecedented data on Saturn and its moons, expanded our understanding of the outer solar system, and demonstrated the potential of collaborative international missions. The success and legacy of Cassini-Huygens continue to inspire and inform future generations of space explorers, reminding us of the limitless possibilities of space exploration and the boundless potential of human ingenuity.

MISSION II: INTERNATIONAL SPACE STATION (ISS)

ENABLED CONTINUOUS HUMAN PRESENCE AND RESEARCH IN SPACE

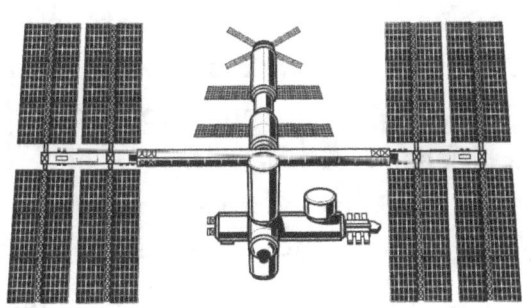

Introduction to the International Space Station:

The International Space Station (ISS) is a symbol of international cooperation and scientific advancement. Launched in 1998, the ISS serves as a microgravity and space environment research laboratory where scientific research is conducted in astrobiology, astronomy, meteorology, physics, and other fields. It is a joint project among space agencies from the United States (NASA), Russia (Roscosmos), Japan (JAXA), Europe (ESA), and Canada (CSA).

Development and Planning:

The ISS project began in the 1980s with the aim of creating a permanent human presence in low Earth orbit. It combined the efforts of several space station projects from the participating countries, including NASA's Freedom, Russia's Mir-2, and Europe's Columbus. The collaboration brought together technology and expertise from around the world, making the ISS a unique and historic venture.

The station consists of pressurized modules, external trusses, solar arrays, and other components. It was assembled in orbit piece by piece, starting with the launch of the Zarya module in 1998. Subsequent launches added laboratories, living quarters, and various research facilities.

Assembly and Launch:

The ISS assembly began with the launch of the Zarya module on November 20, 1998. This was followed by the launch of the Unity module in December 1998, marking the beginning of the ISS assembly in space. Over the next several years, additional modules and components were launched and assembled in orbit through a series of space shuttle missions and Russian launches.

Key modules include:

- **Zarya (1998):** The first module, providing initial propulsion and power.
- **Unity (1998):** The first U.S. module, connecting various sections of the ISS.
- **Zvezda (2000):** The Russian module providing living quarters and life support systems.
- **Destiny (2001):** The primary U.S. research laboratory.
- **Columbus (2008):** The European research laboratory.
- **Kibo (2008):** The Japanese experiment module.

The station's assembly required intricate spacewalks, robotic operations, and international collaboration, with astronauts and cosmonauts working together to complete its construction.

Scientific Research and Discoveries:

The ISS has facilitated a wide range of scientific research, leveraging the unique environment of space. Some of its most significant contributions include:

Human Health Research: Studies on the ISS have provided insights into the effects of long-term spaceflight on the human body, including muscle atrophy, bone loss, and fluid shifts. This research is crucial for future long-duration missions to the Moon, Mars, and beyond.

Microgravity Experiments: The microgravity environment allows scientists to conduct experiments that are not possible on Earth. Research includes fluid dynamics, combustion, materials science, and fundamental physics. These experiments have led to advancements in various fields, including medicine and manufacturing.

Astrobiology and Earth Sciences: The ISS hosts experiments studying the effects of space on biological organisms, contributing to our understanding of life's potential on other planets. Additionally, Earth observation instruments on the ISS provide valuable data on climate change, natural disasters, and environmental changes.

Technology Demonstrations: The ISS serves as a testbed for new technologies, such as advanced robotics, life support systems, and space habitats. These technologies are vital for future exploration missions.

International Collaboration and Achievements:

The ISS stands as a testament to international collaboration. It brings together astronauts, cosmonauts, and researchers from around the world to work towards common goals. This collaboration extends beyond scientific research to include cultural exchanges and diplomatic cooperation.

Current Status and Future Prospects:

The ISS continues to operate and conduct research, with plans to extend its mission through at least 2030. As newer space stations and commercial space habitats are developed, the ISS will serve as a bridge between current low Earth orbit operations and future deep space missions.

The Legacy of the International Space Station:

The ISS has had a profound impact on humanity's ability to live and work in space. It has provided a platform for groundbreaking research, technological advancements, and international cooperation. The legacy of the ISS will continue to inspire future generations of scientists, engineers, and explorers as we push the boundaries of human presence in space.

Personally, I believe the ISS is among the top three missions in the history of space exploration. Its contributions to science, technology, and international cooperation make it a cornerstone of modern space exploration.

In conclusion, the International Space Station is a landmark achievement in space exploration. It has enabled continuous human presence and research in space, provided invaluable scientific data, and demonstrated the power of international collaboration. The success and legacy of the ISS continue to inspire and inform future generations of space explorers, reminding us of the limitless possibilities of space exploration and the boundless potential of human ingenuity.

MISSION 12:
MARS EXPLORATION ROVERS: SPIRIT AND OPPORTUNITY

PROVIDED EVIDENCE OF PAST WATER ON MARS

Introduction to the Mars Exploration Rovers:

Launched by NASA in 2003, the Mars Exploration Rovers, Spirit and Opportunity, were designed to explore the Martian surface and search for signs of past water activity. These twin rovers vastly exceeded their expected lifespans and provided invaluable data on the geology and climate of Mars, significantly enhancing our understanding of the Red Planet.

Development and Planning:

The Mars Exploration Rover (MER) mission was part of NASA's Mars Exploration Program, which aimed to explore Mars and prepare for future human missions. Spirit and Opportunity were built to withstand harsh Martian conditions and conduct scientific experiments on the planet's surface.

Each rover was equipped with a suite of scientific instruments, including panoramic cameras, microscopic imagers, spectrometers, and a rock abrasion tool. These instruments were designed to study the Martian terrain, analyze the composition of rocks and soils, and search for signs of water-related processes.

The Launch and Journey:

Spirit (MER-A) was launched on June 10, 2003, followed by Opportunity (MER-B) on July 7, 2003. Both rovers traveled through space for approximately seven months before arriving at Mars. Spirit landed in Gusev Crater on January 4, 2004, and Opportunity touched down in Meridiani Planum on January 25, 2004.

Scientific Discoveries and Contributions:

Spirit's Discoveries:

Spirit's mission began in Gusev Crater, where it discovered evidence of ancient volcanic activity and the presence of water-altered minerals. One of Spirit's significant findings was the discovery of silica-rich soil, suggesting that hot springs or steam vents once existed in the area. These findings provided strong evidence that liquid water once flowed on Mars, creating conditions that could have supported microbial life.

Opportunity's Discoveries:

Opportunity's landing site in Meridiani Planum proved to be a treasure trove of scientific discoveries. The rover found hematite-rich spherules, known as "blueberries," which formed in the presence of liquid water. Opportunity also discovered cross-bedding in rock formations, indicative of ancient flowing water. These discoveries confirmed that Meridiani Planum once had standing water and possibly a habitable environment.

Longevity and Endurance:

Originally designed for a 90-sol (Martian day) mission, both rovers far exceeded their expected lifespans. Spirit continued to operate until 2010, becoming stuck in soft soil and eventually losing contact. Opportunity, on the other hand, continued to explore Mars for nearly 15 years, covering a distance of 45.16 kilometers (28.06 miles) before a severe dust storm ended its mission in 2018.

Current Status and Legacy:

Although both rovers are no longer operational, their legacy lives on through the vast amount of data they collected. The discoveries made by Spirit and Opportunity have provided a foundation for future Mars missions, including the Mars Science Laboratory Curiosity rover and the Mars 2020 Perseverance rover.

The Legacy of Spirit and Opportunity:

The Mars Exploration Rovers' contributions to science are immense and enduring. They transformed our understanding of the Martian surface, climate, and geology. Their findings provided compelling evidence that Mars once had liquid water, creating conditions that could have supported life.

The success of Spirit and Opportunity demonstrated the effectiveness of robotic exploration in harsh environments and paved the way for future missions. Their legacy continues to inspire scientists, engineers, and space enthusiasts, highlighting the importance of exploration and the quest to understand our place in the universe.

In conclusion, the Mars Exploration Rovers, Spirit and Opportunity, were landmark achievements in space exploration. They provided invaluable data on Mars, confirmed the presence of past water activity, and significantly advanced our knowledge of the Red Planet. The success and legacy of these rovers continue to inspire and inform future generations of space explorers, reminding us of the limitless possibilities of space exploration and the boundless potential of human ingenuity.

MISSION 13:
KEPLER SPACE TELESCOPE

DISCOVERED THOUSANDS OF EXOPLANETS, ENHANCING OUR KNOWLEDGE OF POTENTIALLY HABITABLE WORLDS

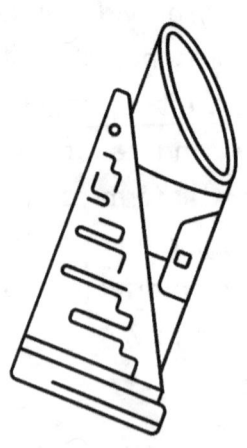

Introduction to the Kepler Space Telescope:

Launched on March 7, 2009, the Kepler Space Telescope revolutionized our understanding of the cosmos by discovering thousands of exoplanets, many of which are potentially habitable. Kepler's mission was to survey a portion of our region of the Milky Way galaxy to discover Earth-sized planets in or near the habitable zone and to estimate how many stars in our galaxy might have such planets.

Development and Planning:

The Kepler mission was named after the Renaissance astronomer Johannes Kepler. The mission was specifically designed to find Earth-sized planets orbiting other stars. The spacecraft was equipped with a photometer that continuously monitored the brightness of over 150,000 stars in a fixed field of view. Kepler's primary goal was to determine the frequency of Earth-like planets in the habitable zones of other stars.

The key instrument on Kepler was a 0.95-meter diameter photometer, which had an array of 42 charge-coupled devices (CCDs). This instrument measured tiny dips in the brightness of stars caused by transiting planets, enabling the detection of exoplanets.

The Launch and Journey:

Kepler was launched from Cape Canaveral Air Force Station in Florida aboard a Delta II rocket. The spacecraft was placed into an Earth-trailing heliocentric orbit, which allowed it to continuously observe the same field of stars without being affected by Earth's shadow or atmosphere.

Scientific Discoveries and Contributions:

Exoplanet Discoveries:
Kepler discovered more than 2,600 confirmed exoplanets and thousands of candidate planets, significantly increasing the known population of exoplanets. Some of its notable discoveries include:

- **Kepler-186f:** The first Earth-sized planet found in the habitable zone of its star, where liquid water could potentially exist.
- **Kepler-22b:** A super-Earth-sized planet in the habitable zone of its star, which was one of the first such discoveries.
- **Kepler-10b:** The first rocky planet discovered outside our solar system.

Statistical Insights:

Kepler's data allowed scientists to estimate that there are billions of Earth-sized planets in the habitable zones of stars within our galaxy. This finding has profound implications for the potential for life elsewhere in the universe.

Multiplicity of Planetary Systems:

Kepler discovered that many stars host multiple planets. Systems such as Kepler-11, which has six confirmed planets, provided valuable insights into the formation and dynamics of planetary systems.

Planetary Characteristics:

The mission provided detailed information about the size, orbit, and composition of many exoplanets, contributing to our understanding of the diversity of planetary systems.

If you are a fan of exoplanets and the search for life outside our Earth, then the Kepler Space Telescope is sure to be your favorite mission. The discoveries it made have captivated the imaginations of scientists and enthusiasts alike, offering a glimpse into the potential for habitable worlds beyond our solar system.

Current Status and Legacy:

Kepler's primary mission ended in 2013 when it lost the second of its four reaction wheels, which are used for precise pointing. However, NASA repurposed the spacecraft for the K2 mission, which utilized the remaining two reaction wheels and thrusters to stabilize the spacecraft and continue its observations in different fields of view. The K2 mission operated until October 2018, when Kepler ran out of fuel and was retired.

The Legacy of the Kepler Space Telescope:

The Kepler Space Telescope has had a profound impact on the field of exoplanet research. Its discoveries have expanded our understanding of planetary systems and the potential for life beyond Earth. Kepler's legacy continues through ongoing analysis of its data and the missions it inspired, such as the Transiting Exoplanet Survey Satellite (TESS), which is building on Kepler's discoveries.

If you are a fan of exoplanets and the search for life outside our Earth, then the Kepler Space Telescope is sure to be your favorite mission. The discoveries it made have captivated the imaginations of scientists and enthusiasts alike, offering a glimpse into the potential for habitable worlds beyond our solar system.

In conclusion, the Kepler Space Telescope was a landmark achievement in space exploration. It provided a wealth of data on exoplanets, significantly advancing our knowledge of potentially habitable worlds. The success and legacy of Kepler continue to inspire and inform future generations of astronomers and space explorers, reminding us of the limitless possibilities of space exploration and the boundless potential of human ingenuity.

For more information on habitable planets, check out our book "20 Habitable Planets: Discovering New Homes for Humanity," which delves into the fascinating discoveries of potentially habitable worlds and what they could mean for the future of humanity.

MISSION 14:
PERSEVERANCE ROVER

LANDED ON MARS TO SEARCH FOR SIGNS OF PAST LIFE AND COLLECT SAMPLES FOR FUTURE RETURN TO EARTH

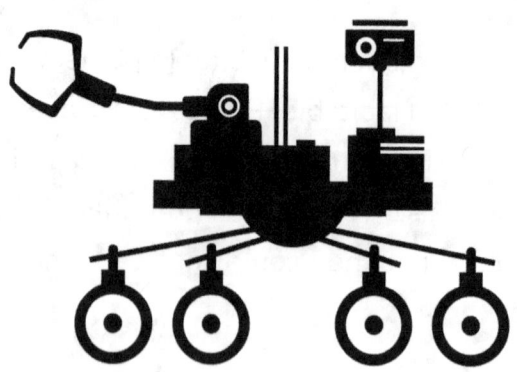

Introduction to the Perseverance Rover:

Launched on July 30, 2020, NASA's Perseverance rover represents the latest and most advanced effort to explore Mars. Part of the Mars 2020 mission, Perseverance's primary objectives are to search for signs of ancient life, collect samples of Martian rock and soil for potential return to Earth, and prepare for future human exploration. This mission builds on the success of previous Mars missions, such as the Mars Exploration Rovers and the Curiosity rover, advancing our understanding of the Red Planet's history and habitability.

Development and Planning:

The Perseverance rover was developed as part of NASA's Mars Exploration Program, with significant contributions from international partners. The rover's design is based on the Curiosity rover, but it includes several upgraded and new instruments to fulfill its scientific goals. Key instruments on Perseverance include:

- **Mastcam-Z:** An advanced camera system capable of high-resolution stereoscopic imaging.
- **SuperCam:** An instrument that provides imaging, chemical composition analysis, and mineralogy.

- **PIXL (Planetary Instrument for X-ray Lithochemistry):** An X-ray fluorescence spectrometer to determine the fine-scale elemental composition of Martian surface materials.
- **SHERLOC (Scanning Habitable Environments with Raman & Luminescence for Organics and Chemicals):** An instrument that uses spectrometry to detect organic compounds and minerals.
- **MOXIE (Mars Oxygen ISRU Experiment):** A technology demonstration to produce oxygen from Martian atmospheric CO_2.

The Launch and Journey:

Perseverance was launched from Cape Canaveral Space Force Station aboard an Atlas V 541 rocket. After a seven-month journey through space, the rover successfully landed on Mars on February 18, 2021, in the Jezero Crater, a site believed to have once hosted a lake and river delta. The landing was executed with the sky-crane landing system, similar to the one used for Curiosity, which allowed for a precise and safe descent.

Scientific Discoveries and Contributions:

Search for Ancient Life:
One of Perseverance's primary goals is to search for signs of past microbial life. The rover is equipped with instruments to analyze the geology and climate of the Jezero Crater, examining rocks and soil for biosignatures that could indicate the presence of ancient life. The rover's SHERLOC and PIXL instruments play crucial roles in this search by providing detailed chemical and mineralogical analyses of Martian samples.

Sample Collection and Caching:
Perseverance is the first rover to carry a system for collecting and caching samples of Martian rock and soil. These samples are stored in sealed tubes that will be left on the Martian surface for a future mission to collect and return to Earth. Analyzing these samples on Earth with advanced laboratory equipment will provide unprecedented insights into Mars' history and potential habitability.

Technology Demonstrations:
The Perseverance mission includes several technology demonstrations to prepare for future human exploration. One of the most notable is

MOXIE, which successfully produced oxygen from Martian atmospheric CO2, demonstrating a critical technology for sustaining human life on Mars. Additionally, the mission carried the Ingenuity helicopter, which made history as the first powered flight on another planet, paving the way for future aerial exploration on Mars.

Environmental and Atmospheric Studies:

Perseverance is also conducting studies of Mars' atmosphere and weather. The rover's MEDA (Mars Environmental Dynamics Analyzer) instrument collects data on temperature, wind speed, humidity, and dust, contributing to our understanding of Martian climate and weather patterns. This information is vital for future human missions and for understanding the potential for past habitability on Mars.

Current Status and Legacy:

Perseverance continues to operate on Mars, conducting its scientific missions and collecting samples. The data and samples gathered by Perseverance are expected to significantly advance our understanding of Mars and its potential for supporting life. The rover's success and ongoing discoveries continue to capture the imagination of scientists and the public, inspiring future missions to Mars and beyond.

The Legacy of the Perseverance Rover:

The Perseverance rover has already made significant contributions to our understanding of Mars, and its legacy will only grow as more data and samples are analyzed. The mission's achievements highlight the importance of robotic exploration in uncovering the secrets of our solar system and preparing for future human missions. Perseverance's success continues to inspire and inform future generations of scientists, engineers, and space enthusiasts, demonstrating the limitless possibilities of space exploration and the boundless potential of human ingenuity.

In conclusion, the Perseverance rover is a landmark achievement in space exploration. It has advanced our knowledge of Mars, demonstrated new technologies for future missions, and provided a foundation for the search for past life on the Red Planet. The success and legacy of Perseverance continue to inspire and inform future generations of space explorers, reminding us of the limitless possibilities of space exploration and the boundless potential of human ingenuity.

If you are interested in Mars colonization and the future of human settlement on the Red Planet, be sure to check out our book, "Mars Colonization: The Red Planet New Era"

The Legacy of the Perseverance Rover:

The Perseverance rover has already made significant contributions to our understanding of Mars, and its legacy will only grow as more data and samples are analyzed. The mission's achievements highlight the importance of robotic exploration in uncovering the secrets of our solar system and preparing for future human missions. Perseverance's success continues to inspire and inform future generations of scientists, engineers, and space enthusiasts, demonstrating the limitless possibilities of space exploration and the boundless potential of human ingenuity.

In conclusion, the Perseverance rover is a landmark achievement in space exploration. It has advanced our knowledge of Mars, demonstrated new technologies for future missions, and provided a foundation for the search for past life on the Red Planet. The success and legacy of Perseverance continue to inspire and inform future generations of space explorers, reminding us of the limitless possibilities of space exploration and the boundless potential of human ingenuity.

If you are interested in Mars colonization and the future of human settlement on the Red Planet, be sure to check out our book, "Mars Colonization: The Red Planet New Era"

MISSION 15:
JAMES WEBB SPACE TELESCOPE

DESIGNED TO STUDY THE EARLY UNIVERSE, STAR FORMATION, AND EXOPLANET ATMOSPHERES

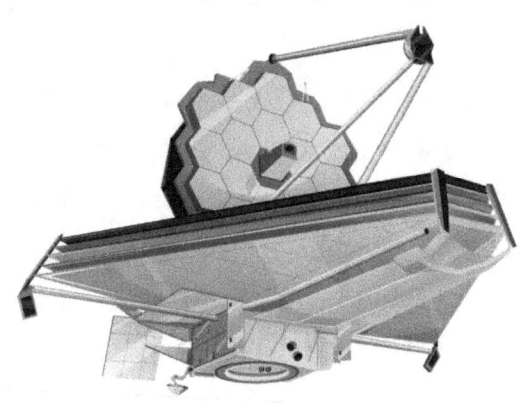

Introduction to the James Webb Space Telescope:

Launched on December 25, 2021, the James Webb Space Telescope (JWST) is the most powerful space telescope ever built. It is designed to study the early universe, star formation, and the atmospheres of exoplanets, providing unprecedented insights into the cosmos. JWST is a collaboration between NASA, the European Space Agency (ESA), and the Canadian Space Agency (CSA), and it builds on the legacy of the Hubble Space Telescope.

Development and Planning:

The development of JWST began in 1996, with the goal of creating a next-generation space telescope capable of observing the universe in the infrared spectrum. This capability allows JWST to peer through dust clouds that obscure visible light, revealing the formation of stars and planetary systems. JWST features a 6.5-meter primary mirror, composed of 18 hexagonal segments made of beryllium and coated with gold to optimize infrared reflection.

The telescope also includes several advanced scientific instruments:

- **Near Infrared Camera (NIRCam):** Captures high-resolution images in the near-infrared range.
- **Near Infrared Spectrograph (NIRSpec):** Analyzes the light from celestial objects to determine their composition and physical properties.
- **Mid-Infrared Instrument (MIRI):** Observes in the mid-infrared range, allowing for the study of cooler objects.
- **Fine Guidance Sensor/Near InfraRed Imager and Slitless Spectrograph (FGS/NIRISS):** Aids in precise pointing and exoplanet studies.

The Launch and Deployment:

JWST was launched aboard an Ariane 5 rocket from the Guiana Space Centre in French Guiana. After its launch, JWST embarked on a month-long journey to its final destination at the second Lagrange point (L2), approximately 1.5 million kilometers (1 million miles) from Earth. At L2, the telescope can maintain a stable position with minimal fuel consumption, allowing for uninterrupted observations.

The deployment process of JWST was complex and meticulously planned. It involved unfolding its sunshield, mirror segments, and other components in space, a process that took several weeks and included hundreds of individual steps.

Scientific Discoveries and Contributions:

Early Universe Studies:

JWST is designed to look back in time to the first galaxies that formed after the Big Bang. By observing these early galaxies, scientists hope to understand the processes that led to the formation of stars, galaxies, and other cosmic structures.

Star and Planet Formation:

JWST's ability to observe in the infrared spectrum allows it to see through dust clouds that obscure young stars and planetary systems. This capability provides insights into the processes of star and planet formation, helping scientists understand the conditions that lead to the development of habitable worlds.

Exoplanet Atmospheres:

One of JWST's primary goals is to study the atmospheres of exoplanets. By analyzing the light that passes through an exoplanet's atmosphere during a transit, JWST can determine the presence of various molecules, including those that might indicate the potential for life. This capability makes JWST a crucial tool in the search for habitable planets beyond our solar system.

Current Status and Legacy:

WST is currently operational and conducting scientific observations. Its early results have already provided stunning images and valuable data, confirming its status as a groundbreaking scientific instrument. JWST's discoveries are expected to reshape our understanding of the universe and answer some of the most profound questions in astronomy.

The Legacy of the James Webb Space Telescope:

The legacy of JWST is still being written, but its impact on the field of astronomy is already evident. The telescope's ability to observe the universe in unprecedented detail will lead to numerous scientific breakthroughs and inspire future generations of astronomers and space enthusiasts.

Personally, I believe the James Webb Space Telescope is among the top three missions in the history of space exploration. Its advanced capabilities and potential for groundbreaking discoveries make it a cornerstone of modern astronomy. Every while, I hear or see something new from JWST, emphasizing its role as the most active mission currently. Just days ago, JWST discovered a new record for the furthest galaxy ever observed, designated JADES-GS-z14-0. This galaxy existed just 290 million years after the Big Bang, providing valuable insights into the early universe. The discovery revealed surprising features, such as the presence of oxygen, indicating that multiple generations of very massive stars had already lived and died in this early galaxy (Space.com) (Geo TV).

In conclusion, the James Webb Space Telescope is a landmark achievement in space exploration. It is designed to study the early universe, star formation, and exoplanet atmospheres, providing unprecedented insights into the cosmos. The success and legacy of JWST continue to inspire and inform future generations of space explorers, reminding us of the limitless possibilities of space exploration and the boundless potential of human ingenuity.

Conclusion

Space exploration has always been one of humanity's most ambitious and awe-inspiring endeavors, pushing the boundaries of our knowledge and capabilities. The missions highlighted in this book, "15 Space Missions That Changed the World," illustrate the remarkable evolution of our journey into the cosmos. From the launch of Sputnik 1, marking the dawn of the space age, to the revolutionary observations of the James Webb Space Telescope, each mission represents a leap forward in our understanding and technological prowess.

These missions have not only expanded our scientific knowledge but have also captured the imagination of millions, inspiring generations of scientists, engineers, and dreamers. The bravery of astronauts venturing into the unknown, the ingenuity of engineers designing and building sophisticated spacecraft, and the dedication of scientists analyzing the data all contribute to the incredible achievements we have witnessed.

The impact of these missions goes beyond the data collected and the discoveries made. They have fostered international collaboration, demonstrated the importance of perseverance and innovation, and highlighted our shared human curiosity. Missions like the International Space Station exemplify the power of global cooperation, bringing together countries from around the world to work towards common scientific goals. The Perseverance rover and the James Webb Space Telescope continue to provide groundbreaking insights, showing that the spirit of exploration is very much alive and thriving.

As we look to the future, the legacy of these missions will continue to inspire and guide us. New technologies and missions will build on the foundations laid by these pioneering efforts, pushing the boundaries of human knowledge and exploration even further. The quest to explore, understand, and eventually inhabit the cosmos is a testament to our innate curiosity and determination.

In conclusion, the story of space exploration is one of courage, innovation, and unyielding curiosity. The missions chronicled in this book represent significant milestones in our journey to understand the universe and our place within it. They remind us of the limitless possibilities of space exploration and the boundless potential of human ingenuity. As we continue to explore the final frontier, we carry with us the lessons and legacies of these remarkable missions, forging a path towards new discoveries and a deeper understanding of the cosmos.

And we would not have reached these heights without the brave souls who left us during missions. May they rest in peace. There is no better way to honor their memories than by continuing the journeys they began, exploring the vast unknowns of space, and expanding the horizons of human knowledge. Their legacy lives on in every step we take into the cosmos, reminding us of the courage and determination that drive us forward.

References

- European Space Agency (ESA). (2024). "James Webb Telescope discovers distant galaxy ever found." Retrieved from Geo News.

- National Aeronautics and Space Administration (NASA). (2023). "James Webb Space Telescope." Retrieved from NASA.

- National Aeronautics and Space Administration (NASA). (2021). "Mars 2020 Perseverance Rover." Retrieved from NASA Mars Exploration.

- National Aeronautics and Space Administration (NASA). (2020). "Kepler and K2 Missions." Retrieved from NASA Kepler.

- National Aeronautics and Space Administration (NASA). (2018). "Mars Exploration Rover Mission." Retrieved from NASA Mars Rovers.

- National Aeronautics and Space Administration (NASA). (2023). "International Space Station." Retrieved from NASA ISS.

- National Aeronautics and Space Administration (NASA). (2017). "Cassini-Huygens Mission Overview." Retrieved from NASA Cassini.

- National Aeronautics and Space Administration (NASA). (2018). "Hubble Space Telescope." Retrieved from NASA Hubble.

- National Aeronautics and Space Administration (NASA). (2024). "Voyager - The Interstellar Mission." Retrieved from NASA Voyager.

- National Aeronautics and Space Administration (NASA). (2021). "Valentina Tereshkova: The First Woman in Space." Retrieved from NASA History.

- National Aeronautics and Space Administration (NASA). (2023). "Apollo 11: First Men on the Moon." Retrieved from NASA Apollo.

- National Aeronautics and Space Administration (NASA). (2024). "Vostok 1: First Human in Space." Retrieved from NASA Human Spaceflight.

- European Space Agency (ESA). (2024). "The Viking Missions to Mars." Retrieved from ESA Viking Missions.

- National Aeronautics and Space Administration (NASA). (2019). "Apollo 8: First Around the Moon." Retrieved from NASA Apollo 8.

- National Aeronautics and Space Administration (NASA). (2018). "Explorer 1 Overview." Retrieved from NASA Explorer.

- National Aeronautics and Space Administration (NASA). (2023). "Sputnik and the Dawn of the Space Age." Retrieved from NASA Sputnik.

www.ingramcontent.com/pod-product-compliance
Lightning Source LLC
Chambersburg PA
CBHW071524220526
45472CB00003B/1136